André Febeliano da Costa

User Experience and Player Experience

André Febeliano da Costa

User Experience and Player Experience

Evaluating and applying to game development

ScienciaScripts

Imprint
Any brand names and product names mentioned in this book are subject to trademark, brand or patent protection and are trademarks or registered trademarks of their respective holders. The use of brand names, product names, common names, trade names, product descriptions etc. even without a particular marking in this work is in no way to be construed to mean that such names may be regarded as unrestricted in respect of trademark and brand protection legislation and could thus be used by anyone.

Cover image: www.ingimage.com

This book is a translation from the original published under ISBN 978-3-330-75850-6.

Publisher:
Sciencia Scripts
is a trademark of
Dodo Books Indian Ocean Ltd. and OmniScriptum S.R.L publishing group

120 High Road, East Finchley, London, N2 9ED, United Kingdom
Str. Armeneasca 28/1, office 1, Chisinau MD-2012, Republic of Moldova, Europe
Printed at: see last page
ISBN: 978-620-8-29711-4

Summary

I dedicate this work to my parents, who have supported me over the years and always encouraged me to pursue my dreams.

Thanks

The main thanks go to my advisor Ricardo Nakamura, my family, my friends, Mariza and all the members of Interlab who contributed to this work.

1 Introduction

Electronic games, compared to other categories of software, have a characteristic in their development: they are concerned with creating experiences for the player. Experiences that simulate real situations, such as playing soccer or riding a horse, are all experiences that games rely on to create hedonic qualities that players will appreciate (SCHELL, 2008).

Electronic games are also digital systems and, like any product, they create an experience in the interaction with their users. This interaction between user and product is studied within the field of Human-Computer Interaction (HCI) involving concepts such as usability and user experience (HASSENZAHL; TRACTINSKY, 2006). Usability, in particular, was very well documented by Nielsen (1993) and has a focus on studying the functionality of digital systems. However, the limitations of usability in working with hedonic concepts allowed the development of the concept of user experience (HASSENZAHL; TRACTINSKY, 2006; LAW et al., 2009) which also studies the functionality of the system, but looks at the interaction between user and product as a whole.

On the other hand, player experience has emerged due to the specificities of games as entertainment systems. Functionality concepts have proved limited in assessing the full scope of interaction with games (NACKE et al., 2009; SANCHEZ; PADILLA-ZEA; GUTIÉRREZ, 2009b). Player experience still has many similarities with user experience, resulting in a confusion of the terms (NACKE et al., 2009; SANCHEZ; PADILLA-ZEA; GUTIÉRREZ, 2009a).

Despite the similarities, player experience and user experience are considered to be conceptually distinct. While player experience is related to the design of experiences through games, user experience is usually related to the evaluation of the experience provided by a product. Consequently, there is a relationship between user experience and product usability, while player experience relates to the concept of gameplay. This does not mean that usability evaluation in games is unnecessary. Incomplete or confusing interfaces have a negative impact on the experience of using the game.

This work aims to trace the relationship between user experience and player experience, looking for similarities and differences between the two. With this, it is hoped to critically evaluate the use of the tool developed for user experience to evaluate games, also considering the player experience and how this evaluation can be included in the game development cycle.

In order to achieve this objective, this paper presents the concepts of user experience and player experience and the experience evaluation methodologies found in the literature. Based on this literature review, two experiments are described which aim to evaluate the player experience, seeking to verify the ability to use evaluation methods within the game development cycle.

2 User experience

Every interaction with a product gives us an experience. This is not only the result of the product's capabilities, but also of how we interact with it. Garrett (2010) presents user experience from this perspective:

"User experience is about how it works on the outside, where a person comes into contact with it." (Garrett (2010))

In other words, when interacting with a product, an experience is generated around it, where the user creates their own opinions about that product. Garrett (2010) seeks to demonstrate that a user's experience when interacting with a product generates an opinion about it in the user. To this end, Garrett (2010) presents experience as a set of user events leading up to the interaction with the product. He argues that these events influence the user's perception of new events, such as interacting with the product. This definition, however, fails to capture the full essence of the UX concept, nor does it provide information on how to evaluate experiences.

User experience is a concept that emerged in the field of Human-Computer Interaction from the limitations found in using usability methods to evaluate interfaces (LAW et al., 2009). It was realized that usability was unable to capture subjective elements relating to the user's experience with the product. The task-based interaction model proved to be incomplete for evaluating all interaction with digital systems with its focus on product quality. Despite the limitations encountered, the adoption of the UX concept demanded discussions in the field about the benefits that UX brings to the area. To this end, the knowledge gathered in other areas has been explored to find elements that help in the study of user experience (HASSENZAHL; TRACTINSKY, 2006).

Based on this, Hassenzahl and Tractinsky (2006) define user experience as a combination of knowledge from different areas, presenting three areas of study that together allow UX to be defined.

The first focus involves the study of interaction beyond instrumental analysis. In other words, studies in this focus seek to analyze product characteristics that do not involve product functionality. However, UX also needs to be concerned with the functionality of the product, since it also impacts interaction.

The second focus of study presented by the authors consists of studying the importance of emotions in computer systems. UX involves the study of emotions and affectivity, always seeking to create positive experiences; studies of this focus seek to detect and resolve negative emotions, such as frustration, that occur during interaction. There is therefore a difference in objectives between studies in this focus and UX studies.

The third focus presented are studies that seek to study the context of technology use, where and when it occurs. In these studies, experience is the combination of the various external factors present, such as the user's emotional state. These studies, however, work on experience in a comprehensive way, while UX consists of the study of experience within the scope of the interaction between user and product.

These three focuses present points of study that are of interest to UX, but none manages to cover UX completely. According to the authors, UX is a consequence of the user's mental state, the characteristics of the system and the context in which the interaction takes place, as Hassenzahl and Tractinsky (2006) put it. For this reason, only the combination of the three focuses makes it possible to understand the interaction.

Because user experience involves so many different areas, its definition has become scattered among the various members of academia and industry who study UX. Garrett (2010) presents user experience as the interaction between user and product, while Hassenzahl and Tractinsky (2006) present user experience as a consequence of the interaction between user and product. However, in both cases we have a very broad definition: there is no definition of how this interaction takes place or a definition of the product with which the user interacts. Law et al. (2009) and Obrist et al. (2012) present this difficulty in defining UX as a consequence of using several different concepts. There is also the fact that experience analysis can take place in a variety of ways, from analyzing just one aspect of a single user's interaction to analyzing all interactions with multiple users.

The advantage of a definition for user experience is that there is a meeting point for all the different views to communicate more clearly. Law et al. (2009) present a survey among members of academia and industry to find the common elements among these various professionals in order to find a base definition. This study is significant because it shows that there are no clear common elements. UX is a dynamic, context-dependent and subjective concept, arising from interaction with the product and which needs to follow user-centered design practices. (LAW et al., 2009) Despite the importance given to UX, the participants did not reach a consensus on a definition for user experience. This result is not surprising, as the participants themselves agreed that user experience is context-dependent and can therefore assume different requirements for different contexts. In other words, the context in which each participant is inserted in the UX study will influence the requirements and definition of UX.

Without consensus, Law et al. (2009) recommend an approach to the study of user experience:

"user experience to be scoped to products, systems, services, and objects that a person interacts with through a user interface." (Law et al. (2009))

This recommendation is in line with Garrett (2010) and Hassenzahl and Tractinsky (2006), but better defines the object with which the user interacts and how this interaction occurs. Despite this, the recommendation fails to indicate how to approach interaction between other users and does not provide a time period for the interaction to occur. Law et al. (2009) argue that experience is something individual and the focus of user experience is to analyze only the individual experience. However, if there is social interaction, it should be noted that the interaction between both users should be analyzed, as it is an integral element of the experience. As for when the experience should be verified, the authors argue that it should be during the interaction between the user and the product, as this is when the user is in direct contact with the product. However, there is an interest in analyzing the experience in the long term, especially for the industry; verifying the experience after the interaction makes it possible to study the aspects that lead to the user's return (LAW et al., 2009). Checking the experience before the interaction takes place is not so interesting for usability, but it is interesting for other areas such as user-centered design, seeking to ensure that when the interaction takes place, it causes the best possible experience.

2.1 Game Development

One of the aims of this work is to understand at what points during the development of games it is possible to carry out tests with users in order to evaluate their experience. Identifying these moments allows us to discuss the relevance and role of these tests in development. However, it is common for the games industry to focus on code quality and development tests (HODENT, 2014). In addition, games have certain characteristics that make user testing difficult, such as the need to avoid early disclosure of game elements like the script.

Digital games are software: digital systems developed for entertainment. Therefore, software development methodology can be used as a basis for game development. With a few restrictions, game development can be imagined as similar to any software where development begins with a survey of requirements, seeking to determine the basic needs that need to be met. The requirements raised at this stage present challenges for development that require specification of the methods, languages, architecture and more that will be used to meet all the challenges. At the end of these two stages, development begins in order to build the specified system, at which stage the system is implemented. Finally, with the development of the system completed, tests are carried out to check that what has been developed meets the specifications, as well as to look for new challenges that have not been met, which generate new requirements, starting the cycle all over again.

Games, however, are playful systems that seek to create experiences for players. In its development, the gathering of requirements takes the form of initial conceptualization, where the game is still poorly defined and ideas can be discussed more freely. From the initial definition, the development moves on to a pre-production stage, in which the ideas are tested using prototypes that seek to find their challenges and possible solutions. What these two stages already show is that game development undergoes filtering, narrowing its scope as development progresses (FULLERTON; SWAIN; HOFFMAN, 2004). This behavior is seen in Figure 1, with the scope and size of the project decreasing as the game nears release. In addition, this also indicates that in game development user testing generally takes place during the pre-production and quality testing stages. However, it is common for the pre-production stage to take place internally without using future players, thus limiting the influence that their opinions have on the development of the game to stages where development is already too advanced to make changes that involve higher stages. And so quality testing focuses on finding problems and programming errors that need attention before release, while new needs and corrections to the experience necessitate starting a new development.

Figure 1 - Stages of game development

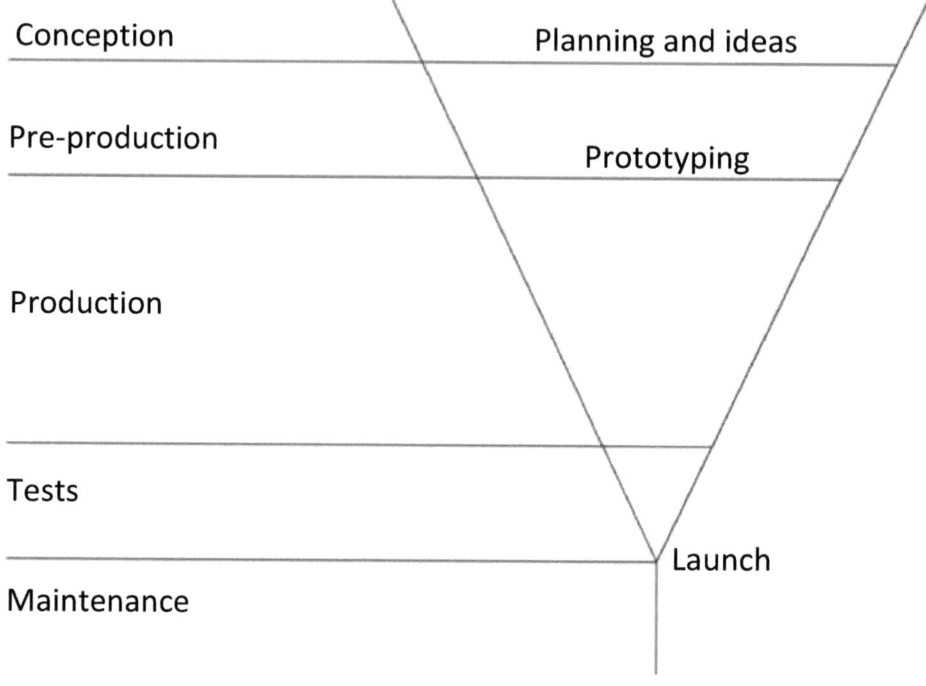

Source: Image adapted from Fullerton, Swain and Hoffman (2004)

2.2 Usability

As previously discussed, user experience is a subjective concept that has gained importance as a result of the limitations found in usability. However, usability has not lost its importance, in fact it plays a major role within the user experience. Usability is an integral part of the product experience because the task that motivated the interaction between user and product must be carried out satisfactorily and this satisfaction in carrying out the task is an integral part of the experience. Therefore, usability problems can impact the user experience and should be checked. (COSTA; NAKAMURA, 2015) It is important to understand what usability is and its metrics, in order to understand what information usability brings to the study of user experience.

Usability has a much narrower focus than user experience. According to Tullis and Albert (2008), usability is the ability of a user to complete the task given to them. This focus on the task and not the interaction is what distinguishes usability from user experience. However, as mentioned earlier, performing the task is part of the interaction, so usability is an important concept for analyzing user experience. Nielsen (1993) presents a more precise definition that is more widely used in the literature. In it, Nielsen (1993) defines usability as an interface property that has five characteristics: *learnability*, *efficiency*, *memorability*, *errors* and *satisfaction*.

Learning is about how easy it is to learn to use the system. This feature is very important, because the first contact a new user has with the product requires them to learn how to use it. In general, what is expected of a system is that it is easy to learn, that the user quickly becomes familiar with the system. However, there are systems that allow for more extensive training in order to reduce the impact of complicated interfaces. This difference is partly due to the complexity and relevance of the product. An information totem in shopping malls is something that should be extremely easy to learn, so that the user spends no time familiarizing themselves and finds their destination quickly. On the other hand, the cockpit of a large passenger plane requires hours of training before the pilot is able to fly the aircraft. However, learning doesn't only occur with new products; updates or similarities with previous versions can help with learning and must be determined. Nielsen (1993) presents the user's success in completing a task in a minimum amount of time using the system as a way of evaluating learning. Looking for the period of time that the user needs to stop learning and start using the system

Efficiency is the characteristic that measures productivity in carrying out the task, for example the reduction in execution time. In this case, it considers the performance of users when they have expertise in using the system, i.e. unlike the previous learning characteristic, this one seeks to evaluate not the beginning of the learning curve but the end of it. To evaluate this feature, therefore, the important thing is to define the level of expertise that users need to display and measure the time they need to complete typical system tasks.

Recall is the ability of the user to use the system again quickly without having to familiarize themselves with the whole system again. This characteristic seeks to evaluate the success of a casual user of the system, who is neither proficient in the system nor a layperson. This user has already used the product and only needs to remember how to use the system. Here, an easy-to-remember and easy-to-identify interface helps returning users to quickly recall prior knowledge acquired from previous uses of the system. This feature is less tested than the others and has two evaluation methods. The first method seeks to carry out a user test by measuring the time required to complete typical tasks, as is also done in learning and efficiency tests, but in this case the user would be someone who is familiar with the system and has not used it for a while. In this way, a more representative test of system recall would be obtained. The second method, on the other hand, consists of working with the user's memory. In this test, the user would carry out a test session with the system and then answer various questions about the system, such as identifying the functions of commands or associating icons with specific commands. Despite being a simpler test to apply, there are systems that try to make the commands the user needs available visually, so the use of memory tests is unnecessary, since *the* system itself reminds the user what they can do.

The few errors feature aims to ensure that the user makes few errors when using the system, that those that do occur can be resolved easily and that none are critical errors. An error in this case would be any action taken by the user that does not achieve the expected result and, therefore, can be easily observed in the number of errors that occur during tests for the other features. It's important to note that errors have different degrees of impact. They can simply be small errors that the user recognizes and corrects, impacting the efficiency of the system. However, catastrophic errors can occur that are not recognized by the user and result in destabilization of the system, affecting the task irrecoverably. These errors need to be identified, counted and resolved.

Satisfaction is the characteristic that comes closest to the concept of user experience and deals with the user's satisfaction when using the system. This characteristic seeks to assess how pleasurable it is to use the system. Studies on this characteristic seek to obtain the user's opinion of the product. This characteristic is even more relevant for systems with a greater focus on entertainment, such as games, as these are systems that the user does not seek to use to solve tasks, but rather to obtain a pleasurable use of the system. Although this is a predominantly subjective characteristic, it is possible to obtain objective data using sensors. However, the use of sensors can make testing more intimidating and, as you want to carry out the tests in conditions that are closer to reality, you need a friendlier atmosphere. Therefore, satisfaction tests mainly involve collecting subjective data by asking users directly for their opinions of the system. These opinions can be obtained using questionnaires using a Likert scale to help standardize and evaluate responses. However, it is important that the questionnaire is tested beforehand to ensure that the questions are understood correctly by the users.

To verify these characteristics, metrics are defined that vary depending on the system being evaluated. According to Tullis and Albert (2008), these metrics allow the interaction between user and system to be verified in terms of effectiveness, efficiency and satisfaction. However, metrics, as Tullis and Albert (2008) state, need to be quantifiable, i.e. obtain quantifiable results that represent the usability characteristics presented by Nielsen (1993), in order to reach conclusions about the quality of the product. However, for the satisfaction characteristic this becomes a limitation, as it is necessary to identify metrics that allow subjective results to be quantified. This subjective nature of the results makes it difficult to replicate the tests for different products and it is necessary to identify specific metrics for each product. In addition, the focus of the usability study on the quality of the task, makes it difficult to analyze aspects of the interaction that occur beyond this focus, external influences such as other users or other stages that use the product are not identified in the study if they do not influence the performance of the task. This need makes the analysis of user experience, through the view provided by usability, limited. However, usability is still an option for obtaining the user's opinion of the product.

2.3 Experience design

The concepts presented above, user experience and usability, allow for the study of the interaction between user and product. However, both have different views: while usability sees the interaction with the product only within the performance of the task, UX seeks to analyze the interaction with the product in all aspects in which it occurs, not only from the performance of the task. However, both focus on the user's interaction with the product and evaluate the outcome of the interaction, with the experience only occurring when the user interacts with the product. Hassenzahl (2010) argues that the study of interaction in the opposite direction, where the product interacts with the user to create the experience, is also significant. For this study, Hassenzahl (2010) presents the concept of Experience Design, which seeks to develop products that create experiences.

Hassenzahl (2010) begins his defense of the concept by presenting a simple conceptual model for experience through interaction with an object. This model seeks to answer three questions: "How?", "What?" and "Why?".

"What?" deals with the interactions that a product can perform, usually linked to the product's functionality. While "How?" deals with the actions that the user will perform, i.e. the user's interaction with the product's interface elements and the context in which the interaction takes place. As you can see, these are two interconnected issues in product development that allow us to define the requirements and characteristics of the product and the combination of these issues with hedonic aspects during the interaction create an "experience"

However, Hassenzahl (2010) argues that this approach misses out on a characteristic of interaction, which is the user's motivation to use the product. Hassenzahl (2010) states that when interacting with a product, the user is not looking for interaction because of the product's functionalities, but because there is a need that the product fulfills. This user motivation is what "Why?" seeks to address and is what experience design, as presented by Hassenzahl (2010), seeks to focus on first in product design and development. Understanding the motivations that lead users to interact with a product helps to identify the needs that the product must meet in order to interact with the user. Understanding the "Why?" indicates a different set of possible interactions in which the product interacts with the user. Therefore, only after determining the "Why?" does experience design seek to answer the other two questions, which deal with the functionality of the product that best recreates the desired experience ("What?") and how best to present this functionality to the user ("How?"). The concept presented, therefore, seeks to develop products that are in tune with the user's motivation by creating products that are pleasurable to use.

2.4 User-centered design

As seen above, there is a concern with designing the product or system with the user's wishes guiding the development. This is not a recent concern, but one that predates the development of the concepts of user experience and experience design. Although both UX and experience design work with users and their interaction with products, both do not include them as a central part of product development.

User experience seeks to identify problems in the product by evaluating the experience during interaction, focusing on all aspects of interaction with the product. Experience design seeks to identify the motivation that leads the user to interact with the product and to develop it based on this motivation, so the focus in this case is on creating a meaningful experience for the user. In both cases, however, the user is not involved during development; their participation is confined to evaluating the solutions developed or in the initial conceptualization.

User-centered design is a concept that predates user experience and experience design (LAW et al., 2009), where the interaction between user and product is not necessarily digital. It is also a concept whose user testing methodologies were the forerunners of user tests aimed at evaluating experience. UCD emerged from an integration of various disciplines, seeking to improve human lifestyles by creating products that are physically, perceptually, cognitively and emotionally intuitive to use (HCDI, 2011). Norman (2006) adds to this definition that UCD seeks to meet the needs and interests of users, concerned with creating products that are easy to use. Garrett (2010) further argues that user-centered design is the practice of creating meaningful and efficient user experiences, where the user is central throughout product development. These definitions therefore show the importance of the user during product development, with a special interest in the user's needs, always seeking to create a product that is meaningful to the user.

Because user-centered design uses the knowledge of various disciplines to understand user needs, various methodologies have been created to obtain this subjective information (LAW et al., 2009). These methodologies have proved capable of verifying not only the user's needs, but some have also made it possible to evaluate the experiences during interaction with the product, also evaluating the chosen design. These methodologies were then used not only in the UCD area but also in interaction areas such as user experience and usability. This allows these areas to obtain data to help evaluate the interaction between user and product for different purposes, such as evaluating new interaction techniques.

3 The evaluation of experience

Experience evaluation appears in the literature in different ways. Depending on the needs of the study, experience evaluation occurs in different ways (OBRIST; ROTO; VAANANEN-VAINIO-MATTILA, 2009). Obrist, Roto and Vaananen-Vainio-Mattila (2009) explore the use of experience evaluations in the literature and present a study that identifies the methods used within different moments of the experience and also during which stages of development. In essence, it presents the evaluation methods used during stages of the product life cycle, from its initial conception to after its launch. With each evaluation seeking to meet the needs that the product presents for each stage of development or regarding the duration of the interaction.

For evaluations seeking to assess experience during specific moments of interaction, Obrist, Roto and Vaananen-Vainio-Mattila (2009) present the methods used at three different moments: emotional evaluation, one-off evaluation and long-term evaluation. Observations are used when the aim is to evaluate the user's emotions or a particular product event. In both cases, the aim is to determine user behavior. For emotion evaluation in particular, measurements can also be applied to tests using auxiliary tools such as sensors, in an attempt to obtain data that goes unnoticed during observation. Finally, the use of personal accounts is used in different ways for the three moments presented.

When checking the use of methods within different stages of development, Obrist, Roto and Vaananen-Vainio-Mattila (2009) show that for three initial stages of development (conceptualization, non-functional prototype and functional prototype), user experience evaluations are carried out using different methods. For tests carried out during the conceptualization stage, visual design and idea evaluation are used. Similarly, the visual design is also of great interest for the non-functional prototype stage, as is the evaluation of the interaction. In both stages of development, these evaluations aim to validate the concepts and ideas of the project in terms of the experience offered. The last stage addressed by Obrist, Roto and Vaananen-Vainio-Mattila (2009) seeks to evaluate functional prototypes during use and therefore, as seen above, seeks to evaluate the emotional experience, both punctually and in the long term. To this end, tests involving prototypes include laboratory tests under controlled conditions, field tests with observations of the prototype in use and market tests using questionnaires with a target audience.

The work presented by Obrist, Roto and Vaananen-Vainio-Mattila (2009) is interesting because it allows us to understand not only the need for specific tests, but also the plurality of objectives that these tests can help with. Furthermore, it is possible to verify their validity by analyzing experience evaluation studies. It is possible to understand that these studies seek to determine the user's experience during some stage of the project's life cycle.

While some studies present a prototype aimed at testing a specific interaction in a preliminary way, whether this interaction is a variation of existing interactions (KNOEDEL; HACHET, 2011; MOERMAN; MARCHAL; GRISONI, 2012), a new interaction (GOH et al., 2014; TORRENTE et al., 2012) or even the simulation of real interactions (LUCERO; HOLOPAINEN; JOKELA, 2011; TSE et al., 2007). Other studies seek to evaluate by observing the user in a laboratory while they are already using the application, usually in a functional prototype. These tests are usually recorded in order to review and capture possible small details that may have been missed during the observation (JOKELA; LUCERO, 2013). The advantage of these studies is that the observation takes place in a controlled environment, without distractions that could influence the results. However, it is also a disadvantage, as it is possible that the user is displaced from their normal environment of use.

A different approach in some studies seeks to minimize the disadvantage mentioned above by carrying out the test in uncontrolled environments, i.e. outside the laboratory (LUOJUS et al., 2013). While this approach makes it possible to see the use of the application in a situation that is closer to the real thing, it is more difficult to record the test without making the experience artificial, making it necessary to adopt other data collection methods in order not to lose observation data (MORRISON et al., 2011). Applications in these studies are already at more advanced stages of development.

The use of questionnaires presented after the application has been used is common in prototype studies. The aim of these is to obtain information directly from the user, such as their opinion and significant moments, or even to express directly which points the user identified with or not during use (LUOJUS et al., 2013; MORRISON et al., 2011).

Experience evaluation presents several valid alternatives, allowing the user experience to be monitored throughout the product's development. However, it is interesting to note that using one method does not preclude the use of others. In fact, there are studies that carry out tests in controlled situations within the laboratory and also carry out tests outside the laboratory, as seen in the work by Luojus et al. (2013). This combination of methods allows for more comprehensive data collection, where the use of one method seeks to cover the disadvantages of another, allowing the researcher to have a broader view of the interaction.

4 Player experience

Player experience is a concept that has various interpretations both in the literature and in the industry. While some try to approach PX as being a player's opinion of the game (PEDERSEN; TOGELIUS; YANNAKAKIS, 2010), where the experience is minimized becoming an indication of a player's feeling when playing. Other studies, as well as those in the industry (HODENT, 2014; WONG, 2014), approach PX as user experience applied to games. However, in the literature there are those who try to approach the gaming experience as a different concept to UX and one that requires specific care.

Schell (2008) argues in his book that game designers need to be concerned about the experience that players will have when playing. What Schell (2008) proposes is that, when interacting with a game, the player engages with the rules, narrative, mechanics and all the other game elements. This engagement is responsible for creating an experience for the player, which developers need to pay attention to. However, Schell (2008) doesn't give a definition of Player Experience (PX), but rather that the experience of playing is an integral part of the player experience, not its totality. In addition, Nacke et al. (2009) present player experience as a concept present in the interaction between game and player. The main difference between these two experiences is that the experience Schell (2008) refers to is integral to the development of the game, while Nacke et al. (2009) present player experience as what the player experiences when playing.

In another approach, Sànchez et al. (2012) also present player experience as the interaction between player and game, but they present characteristics of games that interaction studies should be concerned with when analyzing player experience. The characteristics presented by Sànchez et al. (2012) are similar to Nielsen's (1993) concept of usability, seeking to evaluate the quality of games. Sànchez et al. (2012) seek to relate the study of player experience to the study of user experience by presenting parallels between the concept of usability and their proposed concept of gameplay. Among the characteristics of gameplay are hedonic concepts such as fun, excitement and entertainment. Demonstrating that the recreational nature of games differs from other products and therefore requires a differentiated treatment to analyze the player's experience.

The contributions by Nacke et al. (2009) and Sànchez et al. (2012), however, do not clearly define what player experience is. Both present PX as being present in the interaction between player and game and indicate a relationship of consequence between the interaction and PX. However, stating that player experience is a consequence of interacting with games may be hasty, because among the characteristics presented by Sànchez et al. (2012), entertainment and fun are two subjective characteristics that make up games, forming part of the experience, and are objectives for game development. Therefore, player experience is more than a consequence. By presenting important characteristics for game development, player experience becomes an integral part of game evaluation and is therefore the focus of the interaction between player and game (COSTA; NAKAMURA, 2015).

4.1 Gameplay

Since the player's experience is the focus of the interaction, it is important to define tools and methods for evaluating the game. The concept of *playability* is presented in the literature as a way of evaluating games in relation to the objectives of entertaining, amusing, teaching, among others. However, there is no consensus on a definition for playability. Sànchez, Padilla-Zea and Gutiérrez (2009a) present two strands of study: the first focuses on controls and learning in games and therefore applies the concept of usability to the context of games, while the second presents gameplay as an evaluation of the quality of the game, assessing the challenges, narrative and even the players' emotional connection with the game. Sanchez, Padilla-Zea and Gutiérrez (2009a) present a definition of gameplay that tries to unite these two strands:

"a set of properties that describe the Player Experience using a specific game system whose main objective is to provide enjoyment and entertainment, by being credible and satisfying, when the player plays alone or in company". a set of properties that describe the Player Experience using a specific game system whose main objective is to provide enjoyment and entertainment, in a meaningful and satisfying way, whether the player

But what these properties might be is a challenge, since PX is a concept that encompasses various subjective characteristics. Sànchez, Padilla-Zea and Gutiérrez (2009a) argue that it is necessary to be able to detect the non-functional objectives present in games. The authors argue that gameplay needs to evaluate the game not just as a product but as the sum of the various areas that make it up, including interaction with other players or people, the art of the game, the mechanics and other elements present in the interaction with the game. These are the facets of gameplay as defined by Sanchez, Padilla-Zea and Gutiérrez (2009a).

Seven properties are presented which the authors believe are necessary to describe the player experience and which make up gameplay: *satisfaction, learnability, effectiveness, immersion, motivation, emotion* and *socialization.*

Sànchez, Padilla-Zea and Gutiérrez (2009b) state that these seven properties are capable of verifying the different elements that make up a game and make it possible to assess the quality of the player experience. Satisfaction allows us to check how the player feels when interacting with the game, their opinion of the quality of the game for them. Fun, disappointment and attractiveness are characteristics that we try to observe in this property.

Learning allows us to check the player's development and knowledge of the game, their ability to understand the rules of the game and master them. Characteristics that influence this property are divided into those related to the game itself and those related to the player. Game-related characteristics include the difficulty of the game, the speed with which the game progresses and new mechanics are introduced, and the ability of the game to use its resources to facilitate the player's learning. Among the characteristics of the player that influence learning the game are: previous knowledge of games, which helps a player in the learning curve of subsequent games; the player's skill level, which indicates how they approach the problems presented by the game; and frustration, which arises during learning the game when the player is faced with a challenge they can't complete.

Effectiveness is the property that seeks to determine the investment of time and resources needed to ensure that players have a positive experience. This property helps to verify the game's design in terms of its ability to remain interesting, its mechanics and the game's progress. In other words, the characteristics we are trying to evaluate relate to the structure of the game, its progress and objectives, as well as how much of the game the player explores and how many of the objectives they complete.

Sanchez, Padilla-Zea and Gutiérrez (2009b) characterize the immersion property as the ability of the game to be creative, being the set of several characteristics: awareness of the consequences of decisions in the game and their influence on the progress of the game; presenting a player absorbed in the game, focusing all their attention and skill; realism, seeking to present a world in which the player can believe to be real; dexterity of the player in interacting with the game; socio-cultural proximity between the player and the content presented in the game.

The motivation property seeks to verify the game's ability to keep the player continuously engaged with the game. For this property, the characteristics that are observed seek to identify the game's influence on the player. Encouragement, where the player remains confident of being able to complete future challenges in the game, seeking to reach new goals. Curiosity, which can be created by using optional elements and objectives that encourage the player to seek them out, leading them to interact more with the various elements of the game. Self-improvement, this feature is about the player or their character improving their skills in the game. Diversity, finally, is a feature that seeks to present diverse elements to the player in order to avoid creating monotony in the player.

Emotion is a property that seeks to verify the involuntary, behavioral reactions and feelings that arise from the interaction between player and game. This property therefore seeks to observe the player's reaction to the stimuli created by the game. The player's attitude towards the content presented by the game, i.e. their behavior in the face of the stimuli presented by the game. Finally, we try to observe the sensory appeal of the game by assessing the player's interest and desire for the game.

Finally, socialization tries to verify the impact of social interaction within a group. This property seeks not only to evaluate the player's connection with others, but also their connection with their character in the game. In order to observe this socialization Characteristics that involve both the interaction of the group as a whole with the game and of the player with the group are of significant importance. Social perception is a characteristic that seeks to observe the players' acceptance of social interaction involving the game. Group awareness is another characteristic that seeks to observe the extent to which players understand that they are part of a group and need to work together to achieve the game's objectives. Understanding the level of personal involvement tries to identify how much the player understands that their individual success within the game impacts on the success of the group as a whole. Another characteristic to be observed for this property is sharing, i.e. how the group organizes the sharing of game resources between group members. Finally, the communication and interaction characteristics seek to assess how group members communicate and interact with each other using the means provided by the game. It is also important to identify the types of interaction that games can present depending on their rules: competitive, when players have individual objectives with only one player being victorious, collaborative, when the success of the group is the objective and cooperative, when players have individual objectives but use the help of the group to achieve them.

The properties presented by Sanchez, Padilla-Zea and Gutiérrez (2009a) cover areas capable of understanding the player experience. However, the authors also argue that the analysis should still be separated between the various elements that make up a game, as the number of non-functional elements present in PX makes a complete analysis very complicated. Therefore, dividing the study between the basic elements of a game is a very interesting approach to research, because understanding that the player's experience also involves narrative, audiovisual and social elements.

Sânchez, Padilla-Zea and Gutiérrez (2009a) defend this division of gameplay analysis, where first there is the gameplay intrinsic to the game, which verifies the gameplay based on the game design and its mechanics. Next, mechanical gameplay deals with the game as a product, the implementation of the game. Interactive gameplay deals with interaction with the game, from where the player interacts with the game, its interface and controls. Then there is artistic gameplay, which covers the artistic elements of the game, both visual and audio, and how they are used in the game. Personal gameplay seeks to analyze the impact the game has on the player personally, i.e. what perceptions and feelings result from interacting with the game. Finally, social gameplay makes it possible to assess the behavior and group dynamics that arise when interacting with the game in a group. The sum of the results of the different gameplays analyzed makes it possible to better understand the quality of the gameplay and verify the player experience.

4.2 Player-centered game development

The development of electronic games is currently carried out in a similar way to software development, with the gathering of requirements, definition of the software architecture, development, testing and validation. However, games have elements in their development that are external to software development, such as character design, scenarios, script, storyboard, among others. Ensuring that the player experience defined at the beginning of the game's development is maintained during all stages of development is an important issue, as noticing gameplay problems that negatively impact the experience during development allows for the creation of more rewarding experiences and helps to increase the game's financial return (HODENT, 2014).

Currently, user and game quality tests are only carried out in the final stages of development, in many cases with the players being the developers themselves and their families. These tests are usually carried out by the QA (*Quality Assurance*) stage (HODENT, 2014) which mainly seeks to evaluate the game in relation to the mechanical gameplay facet, identifying programming and interface problems.

Sanchez, Padilla-Zea and Gutiérrez (2009a), on the other hand, argue that by including analysis of the various forms of gameplay during each stage of development, it is possible to guarantee quality by including players throughout the game's development, from the initial requirements to the final tests. What the authors propose is to include gameplay elements during the various stages of development to guarantee the quality of the game in terms of its playability, creating meaningful player experiences.

Initially, the requirements assessment seeks to identify game elements that affect gameplay attributes in each facet of gameplay. This makes it possible to control the influence of each game element on the experience. In the next stage of game design, the authors argue that the introduction of gameplay attributes improves the efficiency and effectiveness of the gameplay in creating the gaming experience. This stage makes it possible to identify which elements of the game need to be tested in the prototype phase. These prototypes make up the third stage of the proposed player-centered design where the prototypes are tested and evaluated, seeking to validate the game design decisions according to gameplay tests and their impact on the player experience, creating feedback for the previous stages that are re-evaluated, allowing the game to only move on to the final production stage, with the inclusion of the finalized elements when validating the gameplay of the game. This proposal by Sànchez, Padilla-Zea and Gutiérrez (2009a) makes it possible to evaluate specific elements of the game, identifying the most relevant gameplay attributes for each stage and which are most significant for the game genre or player profile.

5 The relationship between user experience and player experience

User experience and player experience are two concepts presented earlier that have emerged in order to provide a better understanding of human-computer interaction. Although Hassenzahl and Tractinsky (2006) present a well-structured definition of user experience, it fails to mention important points regarding how interaction takes place. The recommendation by Law et al. (2009) helps and complements this definition by better defining interaction. The combination of these authors allows us to understand what user experience is and forms the basis of this concept for this work.

On the other hand, player experience is still a concept without a clear definition. Nacke et al. (2009) try to present in a similar way to UX that player experience occurs when there is interaction between player and game, but they don't go very far in presenting a definition of the concept. Sànchez et al. (2012) seek to present a relationship between UX and PX, but do not discuss a formal definition for player experience. However, these authors, together with Schell (2008), allow us to understand that player experience should be something to be achieved, the goal of game development. This understanding is the basis of this concept for this work, allowing us to better understand the similarities and differences between the two concepts.

Among the similarities between the two concepts is the need for both to carry out experience analysis. And, as presented above, the analysis of experiences in interactions is generally carried out at different stages of the product's life cycle. Evaluations are used to identify various factors, addressing subjective qualities in order to determine the experience during the interaction. Therefore, there are similarities in the data observed and methods used for user experience and player experience. However,

Table 1 - Comparison between UX and PX

User experience	Player experience
Productivity	Entertainment

Task	Fun
Eliminate errors	Challenges and mistakes
Intuitive	Learning mechanics
Reduce workload	Growing challenge
Posthumous gratuity	Intrinsic gratification
Technology to make it easier	Technology to challenge

Source: Table adapted from Sànchez et al. (2012)

as Table 1 shows, there are differences between them in terms of key points.

The main difference between UX and PX lies in their study objectives. While UX is the consequence of the interaction between user and product and is therefore used to verify the product's hedonic characteristics and their impact on the performance of the task for which the product was developed (HASSENZAHL; TRACTINSKY, 2006; LAW et al., 2009). PX, on the other hand, is the development objective and aims to evaluate the hedonic characteristics of the interaction as a means of evaluating the designed experience (NACKE et al., 2009; SANCHEZ et al., 2012).

This difference between the objectives raises a series of other differences related to these objectives. While UX seeks to optimize the performance of tasks, eliminating errors, presenting an intuitive interaction and reducing the workload. PX, however, seeks fun, creating increasingly complicated challenges that lead to learning from mistakes. Both approaches can also be described in terms of the use of technology: user experience seeks to use technology in a way that makes the user's life easier, allowing the user to accomplish or produce more in the same amount of time. In this case, the user's gratification is due to the increase in their productive capacity, since they can now perform the same task more efficiently. On the other hand, player experience seeks to use technology to challenge, creating challenges and situations that players must overcome. In this case, the player's gratification is intrinsic to the game, i.e. interacting with the game and overcoming its challenges provides the gratification the player needs.

Another difference is where the interaction can take place. While for user experience this interaction can occur at some point in the product's use cycle. For games, however, this interaction has greater weight in the interaction between player and game. However, it doesn't only occur through a user interface; elements of the game, such as art and mechanics, also interact with the player outside the scope of the interface, creating layers of interaction (SANCHEZ; PADILLA-ZEA; GUTIÉRREZ, 2009a). Despite the subtle difference between the two concepts, the fact that games have hedonic requirements, such as entertainment and fun, ensures that the study of the experience during the interaction between the game and the player is more relevant (COSTA; NAKAMURA, 2015). It is important to remember, however, that the player is also a user, where the game is the product with which the player interacts and, therefore, user experience is also present. What separates the concepts are the requirements of the systems and where this interaction takes place, creating specificities that allow PX to better meet the requirements of games (SANCHEZ et al., 2012).

This common link between UX and PX can be seen in the relationship between usability and playability, as presented by Sanchez, Padilla-Zea and Gutiérrez (2009a). The definition presented by Sanchez, Padilla-Zea and Gutiérrez (2009a) argues that because games have subjective requirements, they need attributes that allow interaction to be discussed more clearly. Because usability and its attributes are used to evaluate the performance of the task for which the product was intended, evaluating it in objective terms. Therefore, using only usability to evaluate games is complicated, and it is necessary to define what effectiveness, efficiency and satisfaction are for games (COSTA; NAKAMURA, 2015). What Sànchez, Padilla-Zea and Gutiérrez (2009a) present in gameplay is an adaptation of usability attributes so that they can be evaluated within the context of games.

While efficiency in usability seeks to evaluate the correct performance of the task. Within the context of games, efficiency consists of something close to usability, where a game must present different characteristics, such as a high degree of player engagement, analyzing characteristics such as the amount of gameplay explored and the quality of the game design (SANCHEZ; PADILLA-ZEA; GUTIÉRREZ, 2009a).

Efficiency, on the other hand, is more difficult to define in games; for systems in functional contexts, efficiency consists of analyzing ease of learning and quantity of production. However, in the context of games, efficiency, like usability, presents problems in its objectivity. Sànchez, Padilla-Zea and Gutiérrez (2009a) present the solution of separating efficiency into its two characteristics: necessary effort and productivity, which within gameplay are presented as learning and immersion. These are two concepts within the scope of the product that allow us to evaluate the effort required for the player to learn to play and evolve their learning of the game, as well as allowing us to evaluate the immersion that the game provides, a subjective concept. Satisfaction is the usability attribute that has subjective characteristics, but focuses on satisfaction with the product. While satisfaction with a game is also important, evaluating the game's fun, attractiveness and degree of disappointment, Sanchez, Padilla-Zea and Gutiérrez (2009a) argue that there are elements external to the game that need to be evaluated and that influence player satisfaction. Therefore, according to the authors, gameplay still needs to verify attributes such as motivation, emotion and socialization.

These three attributes influence satisfaction with the game in different ways, motivation verifies the reasons why the player has decided to interact with the game. The main appeal of assessing motivation for games is that these are entertainment-oriented systems, there is no task that requires interaction, interaction occurs entirely because the player has chosen to interact with the game and studying the motivations behind this decision allows us to better understand the satisfaction that the game brings to the player.

The study of the user's emotional responses makes it possible to understand the player's behavior when interacting with the game. This is an attribute that explores the emotions caused by the game in order to associate them with the player's reactions and behaviors. Visual, audio and narrative appeals are best analyzed within this attribute.

Socialization is an attribute presented by Sanchez, Padilla-Zea and Gutiér- rez (2009a) which seeks to verify the social aspect present in games related to other players or characters in the game. The relationship between this aspect and usability satisfaction comes from the influence of social elements, such as interaction with other players, on the player experience. To this end, the authors argue that the player's degree of social engagement with others should be observed. Elements such as how the group is organized around the game, communication between group members and the interaction between the group and the game should be studied. This study makes it possible to verify the impact of the game on group satisfaction.

A final detail that Sanchez, Padilla-Zea and Gutiérrez (2009a) present is that the definition of gameplay presented can only be fully explored if gameplay is studied within all possible interactions, thus resulting in the analysis of the player experience as a whole, involving not only the user interface but also the game mechanics, visual and sound art, and other elements of the game.

The definition of gameplay presented by Sànchez, Padilla-Zea and Gu- tiérrez (2009a) clearly shows the differences between the concepts of user experience and player experience. However, it also allows us to better understand the limitations of using UX within the context of games. This does not mean that studying user experience in games is wrong, but it does show that it is necessary to work on the definition of UX in order to bring it closer to the context of games. Elements such as usability and interaction have limitations in the context of games and need to be adapted to capture the specificities of games (COSTA; NAKAMURA, 2015).

However, player experience still has a lot to build on user experience, especially in relation to game development; UX uses the methods and concepts developed by the field of user-centered design and this connection between the two can provide indications of how to use the concept of PX within game development. Sanchez, Padilla-Zea and Gutiérrez (2009a) present a player-centered game development approach that seeks to bring PX analysis into game development by making it part of the development cycles. Sanchez, Padilla-Zea and Gutiérrez (2009a) argue that gameplay is usually evaluated in the testing periods of games after much of the production has been completed. (HODENT, 2014) The authors argue that an approach close to that used in user-centered design is possible, where development begins with the gathering of gameplay requirements, followed by the design of the game and the definition of the software production system. Development then begins with the development of playable prototypes that are evaluated using the gameplay proposal presented by Sànchez, Padilla-Zea and Gutiérrez (2009a), creating feedback on the requirements defined at the start of development, before moving on to develop the final elements of the game. However, it is important to note that UCD always seeks to identify the user's needs and from these obtain the requirements of the product to be designed, while the proposal presented by Sânchez, Padilla-Zea and Gutiérrez (2009a) carries out tests with players in order to feed the cycle with information pertinent to the player experience observed in order to correct the course of the game design.

Another approach close to user experience is that presented by Hassenzahl (2010), Experience Design, where he argues that product design should start by identifying user needs and motivations in order to create a satisfying and meaningful experience. This concept is interesting because it is similar to what Schell (2008) presents in his proposal to design games based on basic experiences. However, while Schell's (2008) proposal allows for greater creative freedom in terms of the experience to be designed, Hassenzahl (2010) argues that experience design should identify the project requirements based on the user's needs and motivations and seek to meet these requirements first, so that the product accomplishes the task for which it was designed.

6 Experience evaluations

Having concluded the bibliographical survey by presenting and discussing the concepts of user experience and player experience, the question remains of how to apply these results to the process of designing digital games. To this end, two experiments were carried out. Both experiments will present the expected player profile, the organization of the test, the game chosen and the methods used.

The first experiment aims to apply the usability tests and the methods used to obtain the results, in an attempt to ascertain whether the elements of PX could be observed and how these elements are presented. Similarly, shortcomings in the use of the methods could be analyzed in order to propose adjustments for the second test. The second experiment seeks to evaluate the player experience in a prototype game, thus simulating the test application in the context of the project's development. Both experiments are detailed in the rest of this chapter.

Both tests were carried out using similar capacities, in order to avoid influencing the definition of the devices, systems and organization. The device to be used will be an iPad Mini running the iOS 9.3 operating system. Participants will also be provided with headphones to prevent external sounds from interfering with the test. The test will be recorded using a Sony Handycam HDR-XR200V camera, supported on a tripod focusing on the participant, in order to obtain their reactions during the test. The game screen will also be recorded in order to obtain the participant's interaction with the game. For this task, a MacBook computer running the 10.10.5 operating system was used and the recording was made using the QuickTime Player 10.4 program with the device connected via a USB cable

Figure 2 - Organization used in the tests

Source: Author

to the computer. This configuration can be seen in Figure 2.

6.1 Evaluation of commercial gaming experience

As previously stated, this first experiment seeks to apply techniques found in the literature for evaluating experience. In this way, we hope to obtain a set of methodologies that can be used to study player experience and applied to the next test.

The aim of this experiment is to evaluate a testing methodology applied to a game in terms of its ability to identify variations in the player experience when varying the game's controls. Therefore, for this experiment, it is important to initially determine which experiences, or rather, which characteristics of the experience we are trying to evaluate, in order to obtain help from the literature on which techniques are the most appropriate.

And as seen above, user experience evaluation makes great use of observation and questionnaires, usually applied in controlled environments (JOKELA; LUCERO, 2013). However, this experiment does not seek to evaluate the game in terms of its interaction with the user and its ability to entertain them. This experiment is more specific and seeks to verify the experience that the player has when interacting with the game and what aspects impact this experience. For this reason, the concept of gameplay presented by Sànchez et al. (2012) presents an approach to the study of experience that allows us not only to verify which aspects of gameplay impact the player experience, but also to present the concept of gameplay facets, allowing us to create an evaluation of just one facet. This creates a capacity for comparison that can indicate variations in the experience just by making changes to one of the facets. This ability exists because the concept of gameplay facets states that the gameplay of a game consists of the sum of the various existing gameplay facets. Therefore, by keeping the other facets the same between tests, it is possible to conclude that any change in the experience derives from the changes made to the facet chosen to be tested.

Therefore, carrying out an analysis using the concept of gameplay presents itself as a solution capable of verifying the experience and this experiment therefore seeks to assess whether, through the results obtained, it is possible to verify a change in the player experience. When looking at how Sànchez et al. (2012) carry out the evaluation of game experience, it can be seen that, similarly to the UX study, the use of questionnaires and observation is also the most common, as it allows for the collection of the participant's profile, the observation of the interaction while it is taking place and the collection of the participant's opinion. It is important to note that these tests need to take into account the various facets of gameplay (SANCHEZ et al., 2012) that should guide the questionnaires. With this in mind, the interaction gameplay was selected, in the form of the game control, in order to verify changes in the experience under different facets. The choice was made because the player has direct contact with this facet and it also allows the use of a commercial game that has more than one control configuration.

The game chosen for the test, the desired profile of the participants and the questionnaires to be answered by the players will be presented below. This will be followed by a detailed description of how the observation for the test was carried out and which features of the game will be observed during the tests. Finally, the results obtained by the chosen techniques will be presented.

6.1.1 Commercial game - Bastion

The game that is chosen to carry out this experiment needs to feature different controls, but retain its other gameplay elements, such as visual art and sound. Commercial games have well-developed elements and so the use of different controls is a possibility that many games address, either by developing innovative controls, using physical controls or adapting existing controls from previous games.

Currently, interaction on mobile devices continues to be explored with different approaches to controlling action games. Bastion, the game chosen for this experiment, is a successful commercial game developed by Supergiant Games (SUPERGIANTGAMES, 2011) initially for consoles and later adapted for computers and mobile devices. This game was chosen because of its approach to character control. The game features two different control configurations, the first consisting of an adaptation of console controls with a joystick and virtual buttons such as

Figure 3 - Virtual joystick in the Bastion game controller

Joystick control concept commonly used in mobile games. It features command buttons (A, B, C and S), a virtual joystick in the left corner and a pause button (P).

Source: Author based on the controls present in Bastion

shown in Figure 3. The second configuration presents an attempt to create an action game control for touch screens, involving automatic attacks, gestures and a few buttons (Figure 4). In both configurations, the presence of command buttons is intended to allow the player to activate various character actions, such as attacking, rolling and using special moves. However, they differ in terms of which actions the player has access to. While the joystick configuration allows for a greater number of actions, the touch configuration seeks to automate some of these actions in order to eliminate on-screen buttons.

6.1.2 Participant profile

This experiment aims to verify experience evaluation techniques in order to obtain information about the participant's experience of playing. Therefore, there is no need for a very selective participant profile. However, the participant needs to understand the mechanics of the game and have knowledge of the game.

Figure 4 - Bastion touch base control

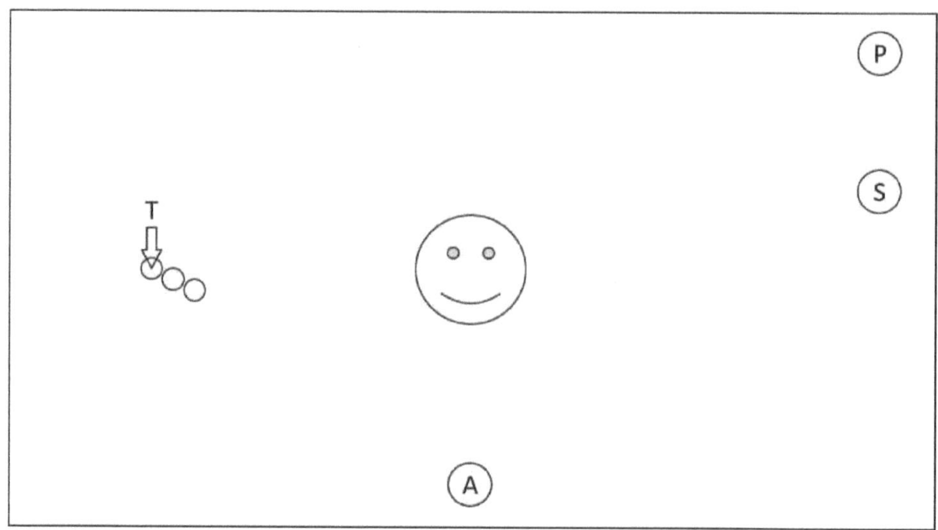

Touch control concept commonly used in mobile games. It features specific command buttons (A and S), the touch position (T) and a pause button (P).

Source: Author based on the controls present in Bastion

s knowledge of using mobile devices. Therefore, the desired profile for this experiment is for the participant to have a good knowledge of English and knowledge of interaction techniques on mobile devices.

6.1.3   Questionnaires

As previously mentioned, the experiment will present two questionnaires to the participants, each for a specific purpose. These questionnaires are necessary in order to obtain data that will help determine the experience that the players witness during the observation. Both questionnaires were designed to help identify characteristics of the gameplay properties presented by Sànchez et al. (2012) that could influence the participants' experience. Characteristics that involve learning the mechanics of the game and emotional reactions are used to verify the quality of the experience and the questionnaires help to obtain more data directly from the participant.

The first questionnaire seeks to identify the profile of the participant, in addition to the desired characteristics of fluency in English and knowledge of interaction techniques on mobile devices, this questionnaire also seeks to find out age, gender and university degree. Although this data is not critical, it is obtained in order to verify the need for additional information for the experience evaluation. To obtain the user's ability to handle mobile devices, a sequence of statements was drawn up with five levels of agreement regarding the use of mobile devices such as *smartphones* and *tablets* (Appendix A). The result obtained will provide simple information on how familiar the participant is with the technology being used in the test.

The second questionnaire, on the other hand, seeks to obtain the participant's opinion of the game. Initially, the participant is asked if they have played the test game before, or if they have played a similar one. Next, the participant is asked to indicate their level of agreement with the game and the control of the game, showing their opinion of the game and the control. At the end, the participant is asked to describe the best and worst moments of the game in their opinion (Appendix B). The aim of this questionnaire is to obtain data that indicates the participant's opinion of the game being tested. So that they can validate the information obtained during the observation.

6.1.4   Note

Bastion is an action game, featuring various game mechanics that are introduced gradually during the game. This experiment will involve the entire tutorial stage of the game, lasting an estimated 20 minutes, during which the participant is invited to verbally express their opinions and actions while playing.

Despite the use of different controls, the mechanics of the game remain similar, as game mechanics are a system of rules with which the player interacts, carrying out actions that are simulated by the system, altering the game and providing feedback to the player, creating a learning process in the player that leads them to carry out a new action (COOK, 2006; KOSTER; WRIGHT, 2004). So, while different controls can lead to different simulations, the game mechanics remain the same, as the rules remain the same, thus allowing for different inputs. Table 6.1.4 below shows the mechanics most influenced by the game control and it is these that will be observed during the test.

Table 2 - Bastion mechanics

	Joystick	Touch	
Player action	Simulation	Simulation	Feedback
Move character	Move in the direction indicated by the joystick	Move to the location selected by touch	Character moves
Attack	Attack with selective weapon - nothing when you touch the button	Automatically attack targets within range with a selective weapon	Character executes attack
Changing weapons	Select another weapon using the button		Weapon selector changing positions
Special attack	Attack with the special ability equipped by clicking on the button		Special attack executed

Table 2 - Mechanics present in Bastion

Joystick		Touch	
Player action	Simulation	Simulation	Feedback
Roll	Scroll in the direction indicated by the joystick when pressing the button	Roll in the direction where a quick double tap is executed	Character executes roll-ment
Interacting with objects	Interact when close by pressing the button	Interact on the move when near	Interaction with objects realizes their functionality

Interact with characters			Interaction with characters starts the con- versation screen
Block enemy attacks	Check for an imminent enemy attack by tapping the shield button In the event of an imminent attack, block and stop the enemy temporarily. Without imminent attack, position shield and, in the event of a subsequent attack, block		Character places shield in front of himself

Table 2 - Mechanics present in Bastion

	Joystick	Touch	
Player action	Simulation	Simulation	Feedback
Execute minutes that I criticize	Execute critical attack by releasing the attack button at the right time	Execute a critical hit by tapping the shield button at the right time	Increased attack damage Power Shot" message is displayed

Table 3 - Initial questionnaire - Bastion

	I disagree		Neutral	I agree	
	Total	Partial		Partial	Total
I use my smartphone every day					4
I play on my smartphone every day	1	1		1	1
I use a tablet every day	2			2	

6.1.5 Results and discussion

Four people took part in this test, aged between 22 and 43, all of whom were fluent in English and owned *smartphones*. However, as Table 3 shows, only two of them play games, and extending this analysis to *tablets* reveals an even greater difference.

Although the results indicate that the participants are not very familiar with games, the results of the first questionnaire were in line with the profile initially desired, with participants fluent in English and knowledgeable about interaction techniques on mobile devices. However, during the tests we realized the importance of the participant's age and habit of playing games. What was observed in the tests was that the older participant had difficulty understanding the game mechanics, a fact that was aggravated by an interruption in the test by the observing researcher when he indicated that the participant could change the controls, and this interruption continued to influence the participant's behavior throughout the rest of the test. On the other hand, the lack of practice on the part of the player was apparent when another participant failed to understand the game mechanics of one of the weapons. In both cases these situations left the participants frustrated, believing that the game removed their control. These failures demonstrated that the methodology needs to be revised in terms of determining the participant's profile and seeking greater familiarity with games.

After the initial questionnaire, the participants were invited to play the game and, as mentioned above, the game mechanics were observed (Table 6.1.4). What was observed was that the large number of mechanics to be observed resulted in some mechanics being used little or not at all. The Special Attack, Roll and Perform Critical Hit mechanics were not performed by any of the participants, which means that there is no variation to the experience, as the participants showed no interest or knowledge in performing these actions. Another little-used mechanic was Block enemy attack, where only one participant used the game mechanic after the tutorial that teaches it. The low adoption of these mechanics made it difficult for all the players to play, as the tactics employed by the participants to advance in the game by not using these mechanics proved to be less efficient.

On the other hand, the other mechanics presented were used successfully, in particular the mechanics of interacting with objects and characters. All the participants quickly identified the objects and characters that allowed them to interact, without any difficulty. The Move Character, Attack and Change Weapon mechanics were used with varying degrees of success. While everyone managed to carry out the command to move the character, either using the virtual joystick or the touch commands, the participants were confused in understanding how the selected control influences the movement action. This confusion led to moments of frustration for the participants when they had difficulty moving the character around the scene as far as they wanted.

That said, what was observed was that the participants who used the virtual joystick control demonstrated a faster understanding of the control and a lower occurrence of movement difficulties. Another game mechanic where the virtual joystick proved to be more significant was in the Attack mechanic, where participants who used the touch control did not demonstrate an understanding that the character automatically attacks when there is a valid target in range of the chosen weapon, with participants at times making game decisions that put them in risky situations in the game and made changing weapons an underused mechanic, as participants preferred the pistols to the bow they receive in the second part of the tutorial. This brings us to the last game mechanic, Switch Weapon, where participants using touch control were reluctant to perform the action even when it was clearly the right strategy, such as an enemy out of reach that needed to be eliminated. When participants switched weapons to the bow, they were irritated by the fact that the character automatically decided to attack a target and this caused the character to slow down until it released the arrow.

Despite the problems with the game mechanics, the post-test questionnaire showed that the participants found the game fun, quite possibly due to the high quality of the visual and sound art, which was highly praised by the participants. As Table 4 shows, the game had satisfactory results and was also considered to be not very irritating, not very boring and not very frustrating. However, when we look at the results for the game control, we see the need for further questioning in order to better understand the participants' opinions. The controls were considered confusing and did not help the game. These results reflect what was observed, where the participants had difficulty understanding all the mechanics of the game and using them. Another property observed was that all the participants were highly immersed in the game, because even after being asked to verbally report their actions and thoughts, the participants stopped doing so after a while. This shows that despite opinions

Table 4 - Final questionnaire - Bastion

	I disagree		Neutral	I agree	
	Total	Partial		Partial	Total
The game was very frustrating	2	1	1		
The game was very annoying	1	2	1		
The game was very boring	2		2		
The game was a lot of fun			1	1	2
Very confusing control	1	1		2	
Uncomfortable control		2	1	1	
Control made the game easier	1	2			1

different from the game, it still remained interesting.

The most significant result of this experiment, however, is the fact that, despite the problems, it was possible to verify a variation in the experience that the participants had when interacting with the same game with different controls. Despite showing that they were having a fun and engaging experience, the participants felt that they were unable to play the game satisfactorily. This indicates that the application of testing methodologies used to evaluate user experience has proved effective in detecting variations in player experience when applied to just one facet of gameplay. This indicates that the next test can work with more variations in controls, provided that the problems with the profile definition, which allowed extreme results in the observation related to characteristics that the game did not seek to meet, are corrected, and also demonstrates that the game to be tested does not need many mechanics or complex mechanics, since the greatest differences in experience were observed in the control of the character's movement.

6.2 Evaluation of experience in a prototype game

As the previous test demonstrated, it is possible to verify experience using the concept of gameplay and its aspects by analyzing a game from one of the facets of gameplay. This experiment, unlike the first, seeks to evaluate this player experience in a game prototype under development, the results of which will be analyzed in the context of a digital game project. This second experiment continues the study of experience under different game controls, again varying the gameplay facet of interaction. In this experiment, the relationship between different game control techniques and the resulting player experience will be evaluated and how this evaluation could enter into game development.

For this experiment, some adaptations, corrections and improvements were made to the evaluation methodology used previously. This is because problems were identified during testing in the first experiment. As the game to be evaluated is at the prototype stage, it is in a pre-production phase in order to improve the use of experience evaluation techniques. Although the game is not in a production phase as discussed by Fullerton, Swain and Hoffman (2004), this experiment allows us to understand the need for testing and its use in more advanced stages of development. The prototype developed for this work will be presented below, followed by a presentation of the methodology used.

6.2.1   Prototype game - Survive

Survive is a digital game prototype developed for this work. The aim of the game is to survive as long as possible by controlling a yellow circle, while green triangles sweep across the screen that need to be avoided for the game to continue. It's a simpler game in terms of quantity

Figure 5 - Unreal 4 schematic for the circle controls

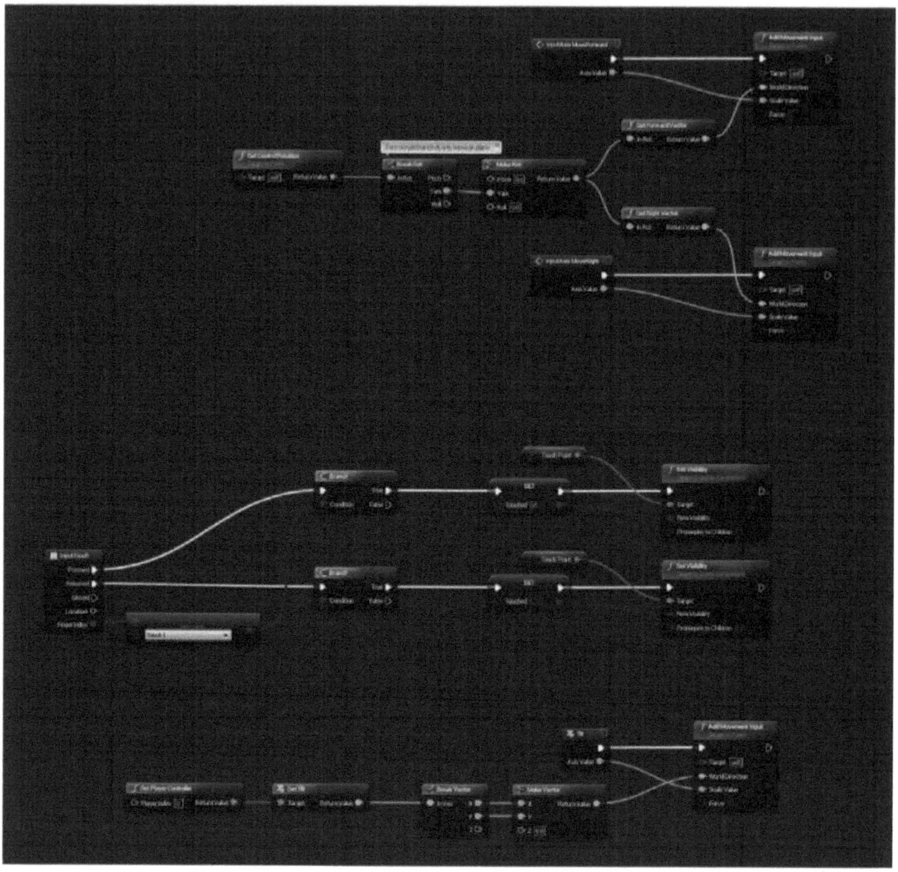

Source: Author

of game mechanics than the Bastion game used in the previous experiment. The game was developed using the Unreal 4.8.3 engine in a language called Scheme, which is a visual programming language (Figure 5). The game was developed for mobile devices, and for this prototype specifically for the iPad Mini device. The game features three different character controls (Figure 6), a virtual joystick (Figure 7), a touch-based one (Figure 8) and the third using the device's gyroscope (Figure 9).

Figure 6 - Controls available in the Survive game

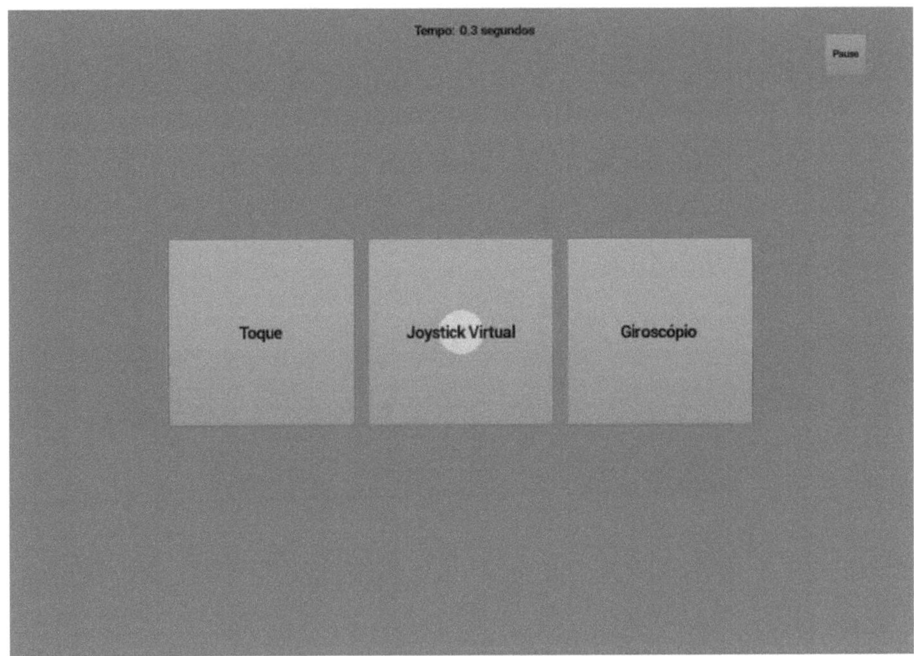

Source: Author

6.2.2 Participant profile

The profile used previously worked as long as we were only hoping to apply the techniques. However, some particularities have occurred that need to be taken into account now and, therefore, information such as age and target audience becomes more important. As this is a game project for mobile devices, the participants must not only be familiar with the touch technology present in these devices, but also be young people between the ages of 20 and 28 studying at university. This age group was chosen because it is the age group of the game's target audience. While the choice of young people studying at university level is due to the fact that these students have greater contact with new technologies and new interaction techniques, as they are located within the academic environment. Fluency in

Figure 7 - Virtual joystick as a control in Survive

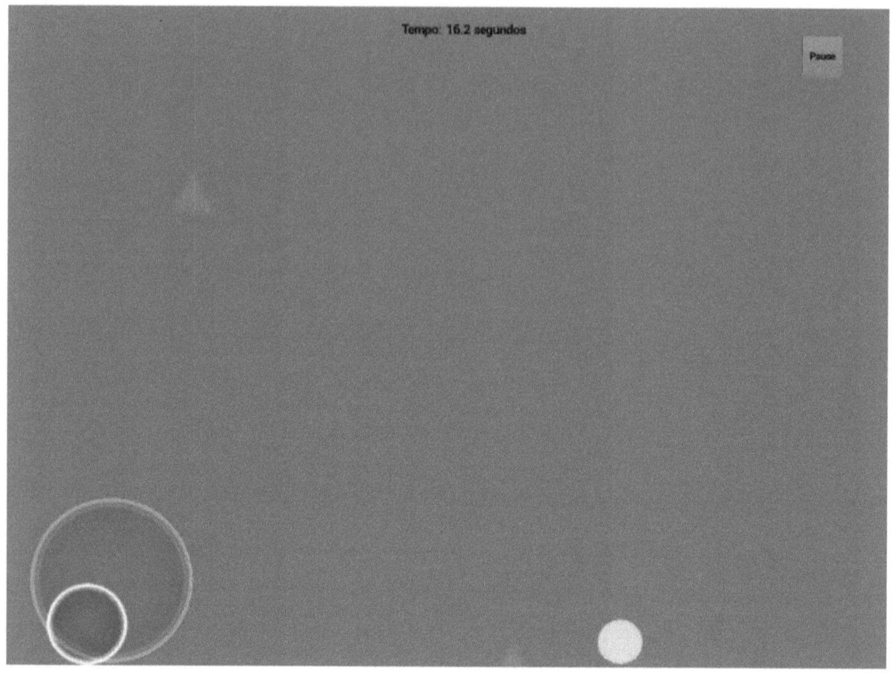

Source: Author

English is not a profile requirement since this time the game is in Portuguese, but it does allow the participant to have greater contact with new interaction techniques made available by the world.

6.2.3   Questionnaires

Similar to the first, this experiment will use two questionnaires. The first will define the participant's profile, while the second will seek to obtain data that indicates the participant's opinion of the game. Despite these changes, the previous questionnaire does not need to be altered and can be reused to obtain the profile data (Appendix A).

The second questionnaire (Appendix C), however, required changes

Figure 8 - Touch-based control in the Survive game

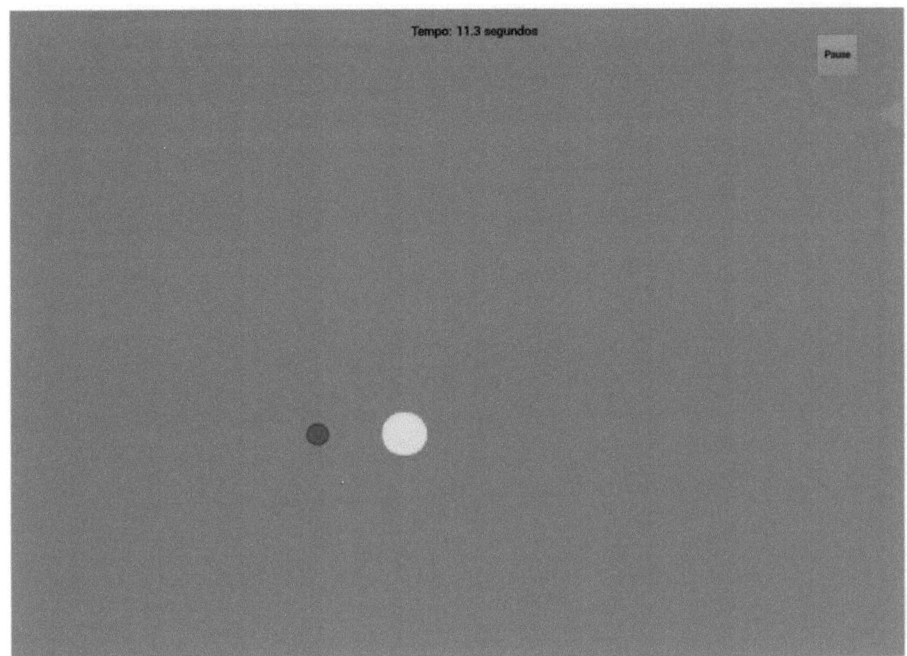

Source: Author

more precisely in its format where the statements about the game and control were presented together. The questionnaire was also designed to obtain more of the participant's opinions about the game. This questionnaire first seeks to find out if the participant has played a similar game before and then the questionnaire seeks to obtain the participant's opinion of the game with a series of statements that the participant must indicate their level of agreement with. The statements are designed to address possible observations during the test and possible emotions that the participant may experience during the test. At the end of these stages, the participant is asked what was the best and worst moment of the game. The same approach is then taken to the game controls.

Figure 9 - Survive game gyro control

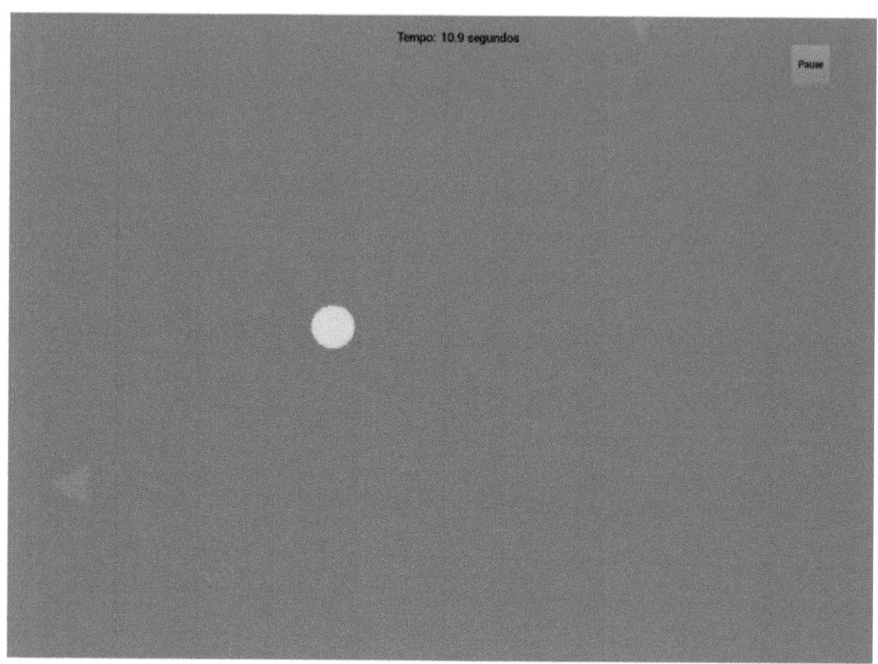

Source: Author

6.2.4 Observation

Survive is a survival game and therefore its matches are short, a common feature of mobile games. Unlike the previous experiment, Survive doesn't have a tutorial teaching the game mechanics and so these will be presented quickly to the participant, as seen in Table 5 below. The test will last 10 minutes, during which time the participant will be informed that they will have to restart the game using the same controller and will also be invited to verbally express their opinions and actions during their interaction with the game.

Table 5 - Mechanics present in Survive

	Joystick	Touch	Gyroscope	
	Simulation	Simulation	Simulation	Feedback
Player action				
Move circle	Move in the direction	Move in the direction of the	Move in the direction indicated	Circle moves

50

	indicated by the selected location by the gi- roscope joystick	
Collide with triangle	Check for collision with the tip of the triangle. If so, the match ends	End of game screen

Table 6 - Initial questionnaire - Survive

	I disagree		Neutral	I agree	
	Total	Partial		Partial	Total
I use my smartphone every day					6
I play on my smartphone every day	1	2		2	1
I use a tablet every day	3		1	1	1
I play on my tablet every day	4		1	1	

6.2.5 Results and discussion

Six people took part in this test, three women and three men, all in the desired age range between 20 and 28 years old, studying at university level and with knowledge of interaction techniques on mobile devices. And as Table 6 shows, for this test half of the participants also play games on mobile devices. This more closed profile proved better at obtaining a set of participants with more similar behaviors when playing.

As this is a game with few game mechanics present, all the participants quickly and clearly understood the rules of the game when they were presented before starting the test. During the tests, the participants had a positive response to the game. Despite the slow start of the game, with few enemies and small movements, the game did not become boring. Participants noticed that the frequency of enemies increased as time progressed and commented that at one point the number of enemies increased rapidly and suddenly, creating a challenge that everyone found interesting. However, there was an increase in frustration and boredom when participants ended up dying too early, before the difficulty suddenly increased. This was probably due to the fact that they knew they would have to do the slow initial stage again before they could start having fun again.

The results of the questionnaire, presented in Table 6, show that the majority of participants find the game fun, but easily frustrating. Table 5 also shows that as the game has few mechanics, the participants were able to understand the rules of the game easily.

While the observations showed similarities in terms of the game mechanics and the participants' learning, this demonstrates that, unlike the first experiment, the defined profile resulted in greater control over learning the game mechanics. When we look at the controls, on the other hand, we see a variation in the preferences of each participant. Table 7 shows more distributed results in terms of the participants' preference for the control they used for the tests. There is a preference for the virtual joystick controls over the others, for example while there are indications that the controls match the game, this is only unanimous for the participants who carried out the tests with the virtual joystick. And even when participants have the opportunity to play a few games using the other controls, the virtual joystick is still considered the most suitable control for the game.

However, the biggest problem with the virtual joystick is the

Table 7 - Final questionnaire - Survive

	I disagree		Neutral	I agree	
	Total	Partial		Partial	Total
The game is very frustrating	1	2		3	
The game is very annoying	2	3		1	
The game is very boring	2	3			1
The game is a lot of fun		1		4	1
The game is very confusing	4	1	1		
Very difficult game rules	6				

Very confusing control	5	1			
Uncomfortable control	3	1	1	1	
Control matches the game			1	1	4
Control makes the game easier			3	1	2
Control gets in the way when playing	3	1	1	1	
Control is easy to learn				1	5
Control is very precise		2		2	2

slow response to fast movements. The touch control, on the other hand, is considered the most precise, but it takes up a lot of screen space with the player's hand moving the circle, causing frustration when there is a collision with a triangle that has appeared in the area hidden by the player's hand. Finally, the gyroscope control is seen as the most interesting of the three and the most sensitive, but the calibration carried out by the device was not sufficient, making the control uncomfortable to use. Because of its sensitivity, the control is perceived as the least precise of the three options. These results demonstrate the importance of the concept of effectiveness in game control, both so that new controls can achieve greater use against well-established controls, and to enable a better experience when using the control.

The experience evaluation of this game, despite its small scope in just the game controls, few mechanics present and limited profile, proved to be of great value, as it helped to identify problematic points in the player experience. Boredom, frustration and irritation appeared as sensations that need to be dosed to avoid causing a negative player experience. The analysis carried out here made it possible to identify that while the game is fun, it has points of interest that need to be re-evaluated in the game design. Boredom in the opening seconds before the difficulty increases needs to be addressed in order to guarantee a positive experience for any player at any skill level. As for the frustration that has been observed, it is necessary to check whether it is directed at the game or at the player themselves. As long as it is directed at the player themselves, this frustration can become a motivation to improve the game, while frustrations directed at the game, such as the fact that the virtual joystick is slow to respond, allow the developer to make adjustments or even re-evaluate the solutions chosen.

Overall, the game presented proved to be a positive experience for the participants. The preference for the virtual joystick showed that this controller is preferred by the game's target audience, probably due to their familiarity with real controllers and other games. However, the gyroscope proved to be a possible solution, although it needs to be adjusted to find the best configuration to provide the best possible experience.

It is important to mention the great value of the tests. The tests carried out made it possible to review not only the validity of using the chosen controls, but also to validate the initial conception of the game. With the information obtained, the design of the game can be revised, generating new design decisions that result in a better experience. This ability to evaluate the game as a whole demonstrates its importance within the development cycle. Although the game tested is still in the prototype phase, it was possible to identify problems and challenges that need to be addressed within the initial concept of the game, such as the time it takes for enemies to appear in Survive. And although analyzing all the content of the tests is very difficult, due to the amount of information that needs to be processed, the data obtained is still invaluable for the design of the game.

By evaluating the player experience within the development of the game, information is obtained that allows the game design to be significantly altered in order to obtain the best player experience. Another advantage of these tests is that, as they consist of evaluating experience, there are tools and solutions that allow them to be evaluated during the game's development and life cycle (OBRIST; ROTO; VÄÄNÄNEN-VAINIO-MATTILA, 2009). As this is the first test of this project, it's difficult to say at what point the experience test for controls should be left aside; one possibility would be when the tests indicate only the need for specific adjustments, such as what is needed to make the gyroscope control more usable.

However, for game design it is interesting to at least evaluate the game under other facets of gameplay in order to understand how the other elements of game design influence the experience. These tests can be carried out focusing only on each facet as it is more defined, seeking to evaluate the facet according to the design decisions relevant to this facet. Such an approach allows the test results to feed the game design to revise these decisions, always seeking the best player experience.

7 Conclusion

We present the concepts of user experience and player experience based on the approaches of different authors. Although it is possible to identify similarities, it is also necessary to observe the particularities that exist in digital games. Since there is a current trend in game development towards experience-oriented game design, the clarity of the concepts and their interrelationships become important in order to allow coherent design practices to be presented, from conception to the evaluation of solutions.

In this sense, the development of the concept of player experience makes it possible to study the specific characteristics present in games, so as to allow a better understanding of the differences between user experience and player experience. This difference is present due to the need for games to be products of entertainment and amusement. Due to this characteristic, functional evaluations based on usability, where the experience is seen as a consequence of the interaction (HASSENZAHL; TRACTINSKY, 2006), have limitations that need to be known in order to prevent these limitations from influencing the results obtained. And although there is a discussion of the concept of experience applied to functional products over their functionality (HASSENZAHL, 2010), this approach is still very young and the focus of product design continues to be on developing products that meet users' hedonic requirements rather than creating varied experiences.

This work presented the concept of playability defined by Nacke et al. (2009) and Sanchez, Padilla-Zea and Gutiérrez (2009a) as a possible approach to the usability limitations that permeate the study of user experience. This concept of playability has demonstrated, in particular, the specific needs in interacting with games, allowing a dialog with the concept of user experience. This opened up a discussion about the applicability of the concept of player experience within game development by identifying possible similarities with the area of user-centered design. This applicability was deepened with the concept presented by Sànchez, Padilla-Zea and Gutiérrez (2009a) where player experience evaluation tests use the concept of gameplay and facets of gameplay to evaluate the experience from the perspective of different game elements, seeking to obtain information and data that will allow game design to advance in the development cycle. This work also presents an understanding between the concepts of user experience and player experience. By stating that the study of player experience focuses on the experience caused and not on the productivity of the game, it is possible to discuss the differences between the concepts in terms of the focus of study.

Finally, this paper presented two experiments that sought to identify and evaluate the player experience using experience evaluation methods. The aim of these tests was to assess the applicability of their use within the game development cycle. The results obtained made it possible to identify gameplay aspects within the game's mechanics that influence the player experience and therefore make it possible not only to identify and evaluate, but also to manipulate the player experience from the initial design of the game. However, due to the amount of data to be analyzed and the work required to analyze all the data, it is important that the tests are not extended so that the development of the game can go ahead and allow the game to be released.

This work has therefore achieved its objective of identifying the differences and similarities between the concepts of experience discussed in this work (UX and PX), as well as demonstrating that the evaluation of player experience makes it possible to obtain significant improvements in game design. By carrying out tests that evaluate game properties during the development cycle, separated into the facets of gameplay presented by Sànchez, Padilla-Zea and Gutiérrez (2009a), it is possible to obtain significant results for the entire development cycle, allowing the game design to be revised as it is developed. Extensions of this work involve more in-depth analysis to answer at what point in development the experience tests should give way to quality tests. Another possible study would be on a current trend of releasing incomplete games in versions still in development, such as the Early Access program of the Steam digital store. Finally, a further development of this issue would be the application of experience tests to more than one development cycle and using many more participants. The evolution of the impact of the data obtained between each cycle could lead to new questions about the development of games focused on the experience provided.

References

COOK, D. *What are game mechanics*. 2006.

Http://www.lostgarden.com/2006/10/what-are-game-mechanics.html.

Last accessed: Jan/2016. Quoted on page 54.

COSTA, A. F.; NAKAMURA, R. User experience and player experience: a discussion on the concepts and their evaluation in the design of digital games. In: *SBC - Proceedings of SBGames 2015*. Teresina, PI, Brazil: SBC, 2015 (SBGames2015), p. 512-517. Cited 5 times on pages 17, 30, 41, 42 and 44.

FULLERTON, T.; SWAIN, C.; HOFFMAN, S. *Game Design Workshop: Designing, Prototyping, and Playtesting Games*. [S.l.]: CMP Books, 2004. ISBN 1578202221. Cited 3 times on pages 16, 17 and 61.

GARRETT, J. J. *The Elements of User Experience: User-Centered Design for the Web and Beyond*. 2nd. ed. Thousand Oaks, CA, USA: New Riders Publishing, 2010. ISBN 9780321683687. Quoted 4 times on pages 11, 13, 14 and 24.

GOH, W. B. et al. The moy framework for collaborative play design in integrated shared and private interactive spaces. In: *Proceedings of the SIGCHI Conference on Human Factors in Computing Systems*. New York, NY, USA: ACM, 2014. (CHI '14), p. 391-400. ISBN 978-1-4503-2473-1. Available at: <http://doi.acm.org/10.1145/2556288.2557104>. Cited on page 26.

HASSENZAHL, M. *The Encyclopedia of Human-Computer Interaction: User Experience and Experience Design*. 2010. Interaction Design Foun- darion. Chapter 3. Available at: <https://www.interaction-design.org/ literature/book/the-encyclopedia-of-human-computer-interaction-2nd-ed/ user-experience-and-experience-design>. Quoted 3 times on pages 22, 45 and 73.

HASSENZAHL, M.; TRACTINSKY, N. User experience - a research agenda. *Behavior & Information Technology*, v. 25, n. 2, p. 91-97, 2006. Disponivel em: <http://dx.doi.org/10.1080/01449290500330331>. Cited 7 times on pages 9, 12, 13,

14, 39, 40 and 73.

HCDI, H. C. D. I. *HCDI - Human Centered Design Institute*. 2011. Last accessed on: Jan/2016. Available at: <http://hcdi.brunel.ac.uk/>. Cited on page 24.

HODENT, C. Developing ux practices at epic games. In: *Game Developer Conference*. Cologne, Germany: [s.n.], 2014 (GDC Europe 2014). Quoted 5 times on pages 15, 29, 35, 36 and 44.

JOKELA, T.; LUCERO, A. A comparative evaluation of touch-based methods to bind mobile devices for collaborative interactions. In: *Proceedings of the SIGCHI Conference on Human Factors in Computing Systems*. New York, NY, USA: ACM, 2013.(CHI '13), p. 3355-3364. ISBN 978-1-4503-1899-0. Available at: <http://doi.acm.org/10.1145/2470654.2466459>. Cited twice on pages 27 and 49.

KNOEDEL, S.; HACHET, M. Multi-touch rst in 2d and 3d spaces: Studying the impact of directness on user performance. *3D User Interfaces (3DUI), 2011 IEEE Symposium*, p. 75-78, 2011. Cited on page 26.

KOSTER, R.; WRIGHT, W. *A Theory of Fun for Game Design*. [S.l.]: Paraglyph Press, 2004. ISBN 1932111972. Quoted on page 54.

LAW, E. L.-C. et al. Understanding, scoping and defining user experience: A survey approach. In: *Proceedings of the SIGCHI Conference on Human Factors in Computing Systems*. New York, NY, USA: ACM, 2009.(CHI '09), p. 719-728. ISBN 978-1-60558-246-7. Available at: <http://doi.acm.org/10.1145/1518701.1518813>. Cited 8 times on pages 9, 11, 13, 14, 23, 24, 39 and 40.

LUCERO, A.; HOLOPAINEN, J.; JOKELA, T. Pass-them-around: Collaborative use of mobile phones for photo sharing. In: *Proceedings of the SIGCHI Conference on Human Factors in Computing Systems*. New York, NY, USA: ACM, 2011.(CHI '11), p. 1787-1796. ISBN 978-1-4503-0228-9. Available at: <http://doi.acm.org/10.1145/1978942.1979201>. Cited on page 26.

LUOJUS, P. et al. Wordster: Collaborative versus competitive gaming using interactive public displays and mobile phones. In: *Proceedings of the 2Nd ACM*

International Symposium on Pervasive Displays. New York, NY, USA: ACM, 2013.(PerDis '13), p. 109-114. ISBN 978-1-4503-2096-2. Available at: <http://doi.acm.org/10.1145/2491568.2491592>. Cited on page 27.

MOERMAN, C.; MARCHAL, D.; GRISONI, L. Drag'n go: Simple and fast navigation in virtual environment. In: BILLINGHURST, M.; JR., J. J. L.; LéCUYER, A. (Ed.). *3DUI*. IEEE, 2012. p. 15-18. ISBN 978-1-4673-1204-2. Available at: <http://dblp.uni-trier.de/db/conf/3dui/3dui2012.html# MoermanMG12>. Cited on page 26.

MORRISON, A. et al. Collaborative use of mobile augmented reality with paper maps. *Computers & Graphics*, v. 35, n. 4, p. 789 - 799,

2011. ISSN 0097-8493. Semantic 3D Media and Content. Available at: <http://www.sciencedirect.com/science/article/pii/S0097849311001129>. Cited on page 27.

NACKE, L. E. et al. Playability and player experience research [panel abstracts]. In: *DiGRA - Proceedings of the 2009 DiGRA International Conference: Breaking New Ground: Innovation in Games, Play, Practice and Theory*. Brunel University, 2009. ISBN ISSN 2342-9666. Available at: <http://www.digra.org/wp-content/uploads/digital-library/09287.44170.pdf>. Cited 6 times on pages 9, 29, 30, 39, 40 and 73.

NIELSEN, J. *Usability Engineering*. San Francisco, CA, USA: Morgan Kaufmann Publishers Inc., 1993. ISBN 0125184050. Quoted 4 times on pages 9, 18, 21 and 30.

NORMAN, D. *O design do dia-a-dia*. Rocco, 2006. ISBN 9788532520838. Available at: <https://books.google.com.br/books?id=8zd8PgAACAAJ>. Quoted on page 24.

OBRIST, M.; ROTO, V.; VAANANEN-VAINIO-MATTILA, K. User experience evaluation: Do you know which method to use? In: *CHI '09 Extended Abstracts on Human Factors in Computing Systems*. New York, NY, USA: ACM, 2009.(CHI EA '09), p. 2763-2766. ISBN 978-1-60558-247-4. Available at: <http://doi.acm.org/10.1145/1520340.1520401>. Cited 3 times on pages 25, 26 and 71.

OBRIST, M. et al. In search of theoretical foundations for ux research and practice. In: *CHI '12 Extended Abstracts on Human Factors in Computing Systems*. New York, NY,

USA: ACM, 2012.(CHI EA '12), p. 1979-1984. ISBN 978-1-4503-1016-1. Available at: <http: //doi.acm.org/10.1145/2212776.2223739>. Cited on page 13.

PEDERSEN, C.; TOGELIUS, J.; YANNAKAKIS, G. Modeling player experience for content creation. *Computational Intel ligence and AI in Games, IEEE Transactions on*, v. 2, n. 1, p. 54-67, March 2010. ISSN 1943-068X. Cited on page 29.

SANCHEZ, J. L. G.; PADILLA-ZEA, N.; GUTIÉRREZ, F. L. From usability to playability: Introduction to player-centered video game development process. In: *Proceedings of the 1st International Conference on Human Centered Design: Held As Part of HCI International 2009*. Berlin, Heidelberg: Springer-Verlag, 2009. (HCD 09), p. 65-74. ISBN 978-3-642-02805-2. Disponivel em: <http://dx.doi.org/10.1007/978-3-642-02806-9_9>. Cited 13 times on pages 9, 31, 34, 35, 36, 41, 42, 43, 44, 45, 73, 74 and 75.

SANCHEZ, J. L. G.; PADILLA-ZEA, N.; GUTIÉRREZ, F. L. Playability: How to identify the player experience in a video game. In: *Proceedings of the 12th IFIP TC 13 International Conference on Human-Computer Interaction: Part I*. Berlin, Heidelberg: Springer-Verlag, 2009. (INTERACT '09), p.

356-359. ISBN 978-3-642-03654-5. Available at: <http://dx.doi.org/10.1007/ 978-3-642-03655-2_39>. Cited 3 times on pages 9, 31 and 32.

SANCHEZ, J. L. G. et al. Playability: Analyzing user experience in video games. *Behav. Inf. Technol.*, Taylor & Francis, Inc., Bristol, PA, USA, v. 31, n. 10, p. 1033-1054, Oct. 2012. ISSN 0144-929X. Disponivel em: <http://dx.doi.org/10.1080/0144929X.2012.710648>. Cited 8 times on pages 29, 30, 39, 40, 41, 49, 50 and 52.

SCHELL, J. *The Art of Game Design: A book of lenses*. Taylor &

Francis, 2008 (Morgan Kaufmann). ISBN 9780123694966. Available at: <http://books.google.com.br/books?id=LP5xOYMjQKQC>. Quoted 4 times on pages 9, 29, 39 and 45.

SUPERGIANTGAMES. *Bastion*. 2011.

Https://www.supergiantgames.com/games/bastion/. Last accessed: Jan/2016. Quoted on page 50.

TORRENTE, J. et al. Preliminary evaluation of three eyes-free interfaces for point-and-click computer games. In: *Proceedings of the 14th international ACM SIGACCESS conference on Computers and accessibility*. New York, NY, USA: ACM, 2012 (ASSETS '12), p. 265-266. ISBN 978-1-4503-1321-6. Available at: <http://doi.acm.org/10.1145/2384916.2384985>. Cited on page 26.

TSE, E. et al. Multimodal multiplayer tabletop gaming. *Comput. Entertain.*, ACM, New York, NY, USA, v. 5, n. 2, p. 1 - 12, apr. 2007. ISSN 1544-3574. Available at: <http://doi.acm.org/10.1145/1279540.1279552>. Cited on page 26.

TULLIS, T.; ALBERT, W. *Measuring the User Experience: Collecting, Analyzing, and Presenting Usability Metrics*. San Francisco, CA, USA: Morgan Kaufmann Publishers Inc., 2008. ISBN 0123735580, 9780123735584. Cited twice on pages 18 and 21.

WONG, K. Designing monument valley: Less game, more experience. In: *Game Developer Conference*. Cologne, Germany: [s.n.], 2014 (GDC Europe 2014). Quoted on page 29.

Appendices

APPENDIX A - Initial questionnaire

Initial questionnaire:

Participant ( ): __

Age: years

Sex: M ☐ F ☐

College: ____

Course: ____

What is your knowledge of English?

None ☐

Basic ☐

Intermediate ☐

Fluent ☐

In the answers below, please indicate how much you agree or disagree with the statements:

I use my smartphone every day:

☐	☐	☐	☐	☐
Strongly disagree	Partially disagree	No opinion	Partially agree	I totally agree

☐	☐	☐	☐	☐
Strongly disagree	Partially disagree	No opinion	Partially agree	I totally agree

I play on my smartphone every day:

☐	☐	☐	☐	☐
Strongly disagree	Partially disagree	No opinion	Partially agree	I totally agree

I use the tablet every day:

☐	☐	☐	☐	☐
Strongly disagree	Partially disagree	No opinion	Partially agree	I totally agree

I play on my tablet every day:

☐ Strongly disagree

☐ Partially disagree

☐ No opinion

☐ Partially agree

☐ I totally agree

APPENDIX B - Final questionnaire - Experiment 1

<u>Final questionnaire:</u>

Participant (_____): ____

Have you ever used the game presented or a similar game for smartphones/tablets?

Yes ☐

No ☐

Have you played games similar to the one presented for smartphones/tablets? (If so, which ones?)

Yes ☐

No ☐

In the answers below, please indicate how much you agree or disagree with the statements:

The game was very frustrating:

☐	☐	☐	☐	☐
Strongly disagree	Partially disagree	Neutral	Partially agree	I totally agree

The game was very annoying:

☐	☐	☐	☐	☐
Strongly disagree	Partially disagree	Neutral	Partially agree	I totally agree

The game was very boring:

☐	☐	☐	☐	☐
Strongly disagree	Partially disagree	Neutral	Partially agree	I totally agree

The game was a lot of fun:

☐	☐	☐	☐	☐
Strongly disagree	Partially disagree	Neutral	Partially agree	I totally agree

The game controls are very confusing:

☐	☐	☐	☐	☐

Strongly disagree   Partially disagree   Neutral   Partially agree   I totally agree

The gamepad is very uncomfortable:

☐   ☐   ☐   ☐   ☐

Strongly disagree   Partially disagree   Neutral   Partially agree   I totally agree

Controlling the game made it much easier:

☐   ☐   ☐   ☐   ☐

Strongly disagree   Partially disagree   Neutral   Partially agree   I totally agree

The game was a lot of fun:

☐   ☐   ☐   ☐   ☐

Strongly disagree   Partially disagree   Neutral   Partially agree   I totally agree

Describe the best moment of the game (Leave blank if there isn't one):

____

____

____

Describe the worst moment of the game (Leave blank if there isn't one):

____

____

____

Adverse comments:

____

____

____

APPENDIX C - Final questionnaire -

Experiment 2

<u>Final questionnaire:</u>

Participant (_____): ____

Have you played games similar to the one presented for smartphones/tablets? (If so, which ones?)

Yes ☐

No ☐

In the answers below, please indicate how much you agree or disagree with the statements:

The game is very frustrating:

☐ ☐ ☐ ☐ ☐

Strongly disagree   Partially disagree   Neutral   Partially agree   I totally agree

The game is very annoying:

☐ ☐ ☐ ☐ ☐

Strongly disagree   Partially disagree   Neutral   Partially agree   I totally agree

The game is very boring:

☐ ☐ ☐ ☐ ☐

Strongly disagree   Partially disagree   Neutral   Partially agree   I totally agree

The game is a lot of fun:

☐ ☐ ☐ ☐ ☐

Strongly disagree   Partially disagree   Neutral   Partially agree   I totally agree

The game is very confusing:

☐ ☐ ☐ ☐ ☐

Strongly disagree   Partially disagree   Neutral   Partially agree   I totally agree

The rules of the game are very difficult to understand:

<table>
<tr><td>☐</td><td>☐</td><td>☐</td><td>☐</td><td>☐</td></tr>
<tr><td>Strongly disagree</td><td>Partially disagree</td><td>Neutral</td><td>Partially agree</td><td>I totally agree</td></tr>
</table>

Participant (_____): ____

In the answers below, please indicate how much you agree or disagree with the statements:

The game controls are very confusing:

<table>
<tr><td>☐</td><td>☐</td><td>☐</td><td>☐</td><td>☐</td></tr>
<tr><td>Strongly disagree</td><td>Partially disagree</td><td>Neutral</td><td>Partially agree</td><td>I totally agree</td></tr>
</table>

The gamepad is very uncomfortable:

<table>
<tr><td>☐</td><td>☐</td><td>☐</td><td>☐</td><td>☐</td></tr>
<tr><td>Strongly disagree</td><td>Partially disagree</td><td>Neutral</td><td>Partially agree</td><td>I totally agree</td></tr>
</table>

The gamepad matches the game very well:

<table>
<tr><td>☐</td><td>☐</td><td>☐</td><td>☐</td><td>☐</td></tr>
<tr><td>Strongly disagree</td><td>Partially disagree</td><td>Neutral</td><td>Partially agree</td><td>I totally agree</td></tr>
</table>

The gamepad makes it much easier to play:

<table>
<tr><td>☐</td><td>☐</td><td>☐</td><td>☐</td><td>☐</td></tr>
<tr><td>Strongly disagree</td><td>Partially disagree</td><td>Neutral</td><td>Partially agree</td><td>I totally agree</td></tr>
</table>

The gamepad gets in the way a lot when playing:

<table>
<tr><td>☐</td><td>☐</td><td>☐</td><td>☐</td><td>☐</td></tr>
<tr><td>Strongly disagree</td><td>Partially disagree</td><td>Neutral</td><td>Partially agree</td><td>I totally agree</td></tr>
</table>

The game controls are very easy to learn:

<table>
<tr><td>☐</td><td>☐</td><td>☐</td><td>☐</td><td>☐</td></tr>
<tr><td>Strongly disagree</td><td>Partially disagree</td><td>Neutral</td><td>Partially agree</td><td>I totally agree</td></tr>
</table>

The control of the game is very precise:

<table>
<tr><td>☐</td><td>☐</td><td>☐</td><td>☐</td><td>☐</td></tr>
<tr><td>Strongly disagree</td><td>Partially disagree</td><td>Neutral</td><td>Partially agree</td><td>I totally agree</td></tr>
</table>

Describe the best moment of the game (Leave blank if there isn't one):

____

____
70
____

Describe the worst moment of the game (Leave blank if there isn't one):

____

____

____

Describe the best part of the control (Leave blank if there isn't one):

____

____

____

Describe the worst of the control (Leave blank if there is none):

____

____

____

Adverse comments:

____

____

____

I want morebooks!

Buy your books fast and straightforward online - at one of world's fastest growing online book stores! Environmentally sound due to Print-on-Demand technologies.

Buy your books online at
www.morebooks.shop

Kaufen Sie Ihre Bücher schnell und unkompliziert online – auf einer der am schnellsten wachsenden Buchhandelsplattformen weltweit! Dank Print-On-Demand umwelt- und ressourcenschonend produziert.

Bücher schneller online kaufen
www.morebooks.shop

Printed by Books on Demand GmbH, Norderstedt / Germany